Ocean
Pollution

by Maria Talen

LUCENT
B·O·O·K·S

LUCENT *Overview Series*

OUR ENDANGERED PLANET

Library of Congress Cataloging-in-Publication Data

Talen, Maria, 1923-
    Ocean pollution / by Maria Talen
        p.   cm. — (Lucent overview series. Our endangered planet)
    Includes bibliographical references and index.
    Summary: Discusses the relationship of humans and the oceans and
the present pollution problems.
    ISBN 1-56006-104-9
    1. Marine pollution—Juvenile literature.   [1. Marine pollution.
2. Pollution.]   I. Title. II. Series.
GC1085.T35   1991
363.73'94'09162 — dc20   91-15567

To my beloved sons, two of the most informative
of all my great teachers.

The editor wishes to thank Carol Sowell for her valuable contribution
to this book.

Look for these and other books in the Lucent Overview series:

Contents

Introduction

THE OCEANS HAVE been part of the world much longer than human beings have, and human life depends on them in many ways. The oceans help keep the temperature of the earth at reasonable levels by storing and transporting heat from the sun. They are a source of minerals and of about sixty million tons of food for humans each year. Many scientists believe the ocean is the place where all life began.

Marine life, that is, plants and animals living in the ocean, are thought to have existed at least three billion years ago. Eventually these first one-celled organisms probably evolved into more complex organisms. Some of these more complex organisms probably then developed the ability to live on land.

These ancient organisms formed a part of the balanced communities of plants and animals called ecosystems. All the living things in an ecosystem rely on each other. When one or more of the elements in an ecosystem is damaged, the whole system is thrown out of balance. Just as all living plants and animals do, the ancient creatures in the ocean ecosystem produced waste, or substances for which they had no use. The oceans absorbed and processed, or recycled, this waste. Within this ecosystem, in which there was no unusable waste or garbage, ocean plants and ani-

(opposite page) Most scientists believe all life began in the ocean.

7

mals thrived and grew. Later, even the food scraps and body wastes of sparse human populations were absorbed and recycled by ocean communities.

Expansion and increased waste

However, as the human population expanded and became more technologically advanced, its waste products became more abundant and more complex. Humans began to combine natural substances to produce new building materials. The chemicals in these substances blended to create new products whose waste was not easily broken down and processed by the oceans.

By the late nineteenth century, machines were doing much of the work formerly done by humans. Industrial production was taking place around the world, creating many complex products. While industrial production gave people labor-saving and time-saving devices, it also created many by-products for which people had no use. As industries grew and the manufacturing processes became more complex, technology expanded and relied more heavily on chemicals and metals.

By this time, the world's oceans were receiving tons of human and industrial wastes. Thinking they were "throwing away" these wastes, people dumped unwanted material and production refuse onto remote land areas or into rivers, lakes, bays, oceans, or seas. There this waste soon disappeared under the water, and people thought they had gotten rid of it.

But many of the wastes created by agricultural and industrial production contained unnatural and even poisonous substances. These substances could not be recycled into ocean or land ecosystems. The undersea communities began to lose their balance, and some actually died. Crippled by this foreign, harmful matter, they could no longer support new life as efficiently as they once had.

Since people could not see the buildup of waste pollution, they did not realize this buildup was occurring. They were unable to dive to the depths of the seas to examine how the ocean ecosystems worked and how these systems were being changed.

The new science of oceanography

Scientists have been studying and measuring the motions and contents of the sea since at least the middle of the nineteenth century. This study is the science of oceanography. A related science, marine biology, is the study of the animal and plant life in the sea. During World War II, new oceanographic technical instruments were developed that could be used to locate submarines, predict waves, and gather data on the speed and direction of water flow. With this new technology, people were able to study the underwater re-

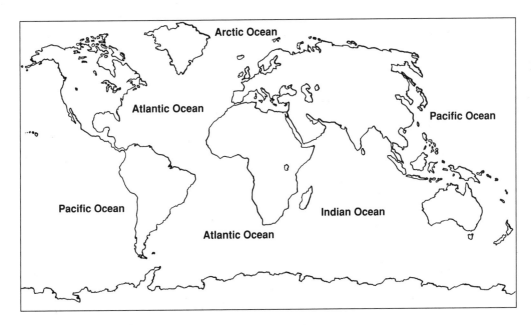

Plastics and other waste products have harmed the oceans and the sea life that depends on them.

gions of the oceans more thoroughly. Only then did researchers realize how little they had really known in the past.

The modern oceanographers soon learned that the oceans, especially the areas near the coasts, had been injured by human waste disposal. Some people found it hard to understand how such vast bodies of water could be seriously damaged or destroyed. People had always believed the oceans were enormous, powerful, and indestructible, and thought they would remain the same forever. Industries and human communities had come to depend on the oceans as waste dumps. They had no idea how they would dispose of waste products without them.

But today we know that many seas and parts of the ocean have indeed been harmed. In fact, some have been destroyed and can no longer support life of any kind. Some species of fish, plants, and animals that have lived in the bays and oceans for centuries have disappeared altogether.

The survival of the oceans

Today many concerned scientists warn that the oceans have reached their limits and cannot survive with continued human waste disposal. Other scientists disagree. They admit that the waters near coasts are overburdened and that dumping should no longer be allowed there. But they insist that the unproductive deep ocean areas, many miles from any inhabited land, are perfect dump sites.

More and more scientists now say that until more is known about the undersea life of the oceans, it would be unwise to continue to dump wastes into any part of the seas. Throughout the world, scientists are doing research to find the safest and best ways to repair existing ocean damage and to make sure the oceans will be clean and productive for future generations.

1

The Importance of the Oceans

THE OCEANS COVER almost 140 million square miles of the earth's surface, an area nearly three times as large as the land areas combined. Within them, the oceans hold fragile, carefully balanced ecosystems, or communities of living plants and animals that depend on each other for life. Because of this dependence on one another to survive, scientists say, damage to any part of the community hurts another part, throwing the system off balance.

The ocean's ecosystems

Why is it so important to maintain the balance among plants and animals that live in the ocean? The ocean ecosystems are part of the larger, inter-locked ecosystem that makes up the entire earth. The earth's environment is like a big machine with many connected parts including the air and ocean currents, which circulate materials from one section of the world to another.

All life on earth depends on the oceans because they are the largest manufacturers of oxygen on the planet. Most living creatures must have oxygen to live, and without the oceans, there would not be enough of this vital gas. Ecosystems on the

(opposite page) The oceans contain delicate, carefully balanced ecosystems.

land also produce oxygen, but these systems are too small to manufacture enough oxygen to sustain all of earth's plants, animals, and humans.

The ocean ecosystem begins with trillions of microscopic plants called phytoplankton. These plants, which float in the ocean, consume sunlight, carbon dioxide from the air, and chemical nutrients from the water. They combine these elements in a process known as photosynthesis, which produces the energy the plants need to live. During photosynthesis, the phytoplankton release oxygen as a by-product.

The food chain

The ocean's ecosystem is also an important part of the world's food chain. The food chain is a natural hierarchy of plants and animals, structured like a ladder, with each rung representing a specific group of plants or animals. Above the phytoplankton in the food chain are zooplankton, microscopic animals that eat the phytoplankton.

Industrial chemical wastes gush into a waterway connected to an ocean. Such wastes threaten the ocean's delicate ecosystem.

The zooplankton are in turn eaten by small fish and other small ocean creatures, which are then eaten by larger fish. These fish provide food for the seals, whales, dolphins, and birds. Eventually some of these fish, mammals, and birds become food for animals that live on land, or for human beings, who are at the top of the food chain.

When a part of the ecosystem of the ocean is thrown out of balance, the plants no longer can photosynthesize efficiently. The system will not produce the oxygen necessary to sustain life in that ecosystem or to support life on land. Eventually the phytoplankton, other bottom-dwelling plants, fish, shellfish, bacteria, and other sea creatures will be unable to reproduce or even to live. Then the higher animals will not have enough food.

Damage to the undersea ecosystem is dangerous to life on land, including humans, in another way. The poisonous chemicals being dumped into the oceans are often absorbed into the bodies of fish, shellfish, turtles, and other creatures that live in or near the ocean. Besides destroying those creatures, these harmful materials can be passed on to land animals and humans who eat fish or

Members of Greenpeace protest ocean dumping.

drink water that contains them.

Human beings discharge hundreds of thousands of tons of waste into the oceans every day. This pollution takes the form of raw and treated sewage, garbage, industrial wastes, deadly chemicals and heavy metals, radioactive waste, and oil. Each kind of waste can create its own type of danger.

Raw sewage spreads disease to swimmers and those who eat contaminated seafood. Large quantities of sludge, or treated sewage, smother bottom-dwelling plants and animals, disrupting the ecosystem. Concentrated industrial wastes are poisonous or toxic to most forms of life. In humans, they may cause cancer, crippling nerve and muscle conditions, and genetic damage that leads to diseases and can be passed on to children. Radioactive wastes can cause death and many serious illnesses in a very short time. A single oil spill can destroy millions of fish and animals.

In addition to their vital functions in sustaining the balance of life on earth, the oceans are important to people for other reasons. In literature, poetry, art, religion, and legends, the oceans have stood for powerful, mysterious forces of nature. The ocean challenges our imaginations with its magnificent creatures such as whales that may weigh as much as 150 tons, dolphins with their charm and intelligence, sharks with huge, powerful teeth, and at least thirty thousand species of wondrously colored and varied fish.

People identify the ocean with Mother Nature or some formidable goddess. Sailors and those who live on the seacoasts have been enthralled by the ocean's power to support massive vessels, its fury during storms, its hypnotic serenity when calm, its unpredictable changes in current. The tides of the ocean are as much a part of nature's rhythm as the sun's rising and setting each day. People go to the ocean to meditate and relax.

In addition, the oceans and seas have become important sources of economic livelihood. Many maritime industries throughout the world including fishing, shipbuilding, transportation, and recreation depend on the seas. The oceans are also the highways for commercial and passenger ships that sail around the world and the means by which the people from one continent first discovered others.

Thus, when people damage the life of the oceans, they damage themselves in economic, spiritual, and physical ways. As humanity reaches the end of the twentieth century, scientists are discovering numerous examples of all these types of harm, including human deaths, caused by ocean pollution.

The Minimata tragedy

One such example unfolded on a small bay on the southwestern shore of the Japanese island of Kyushu. The city of Minimata is located there,

The fishing industry is one of many industries that depends on the oceans for survival.

surrounded by villages where fishing was the main source of income. Once a rustic resort area known for its scenic beauty and healthy climate, today Minimata is much better known as the site of one of the world's most tragic cases of industrial pollution.

In 1952, cats in Minimata began having violent convulsions, then dying. The cats acted as if they were crazy, and some jumped into the sea and drowned. People in Minimata called this behavior "cat dancing" disease.

Soon the same strange illness began to affect some people in the fishing villages. They experienced convulsions and extreme pain. Some developed twisted and crippled arms and legs; others had severe nerve disease, numbness in the limbs or lips, poor coordination, or slurred speech. Some of the people went insane, some became blind, and some died.

After long investigation, scientists found that the cats and the people were suffering from mercury poisoning. Deadly mercury had built up in the flesh of the fish and shellfish they ate, and it penetrated the victims' central nervous systems and brains.

Victims of poisoning

The people of Minimata had been eating fish from Minimata Bay for centuries. Why did the fish suddenly become poisonous to them? Gradually, investigators realized that the deadly substance was methyl mercury, a waste product from a manufacturing process. The Shin-Hihon Chisso Hiryo Company had opened in 1932, making chemical fertilizers. It had piped its wastes, containing the chemical mercury, into the bay since that time. Disposed of at first in modest amounts, this mercury was relatively safe. But over time, the situation changed drastically.

By 1967, scientists figured out that enough mercury had accumulated in the bay area to combine with organic wastes in the water to produce a new compound called methyl mercury, which was highly dangerous to animals and people. This happened as a result of disposing of wastes in the ocean. It was a situation that no one foresaw. It took fifteen years for scientists to trace this development to the factory because the company refused to give information about its manufacturing processes or to submit wastewater for analysis. All the while, the Chisso company continued dumping its wastes into the bay.

In the meantime, at least 155 people near Minimata Bay died from methyl mercury poisoning, and more than one thousand people of all ages suffered effects including permanent deformities and brain damage. Cats, dogs, pigs, crows, crabs, fish, and shellfish died. And human babies were

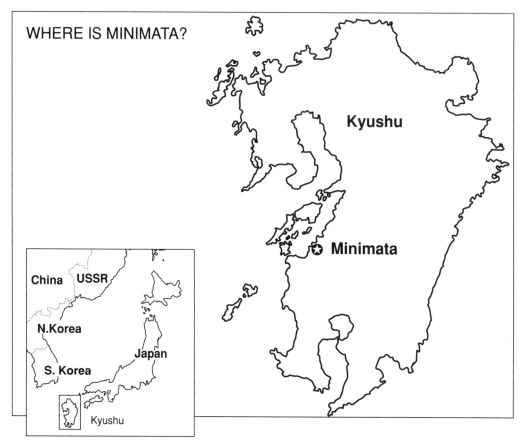

WHERE IS MINIMATA?

Kyushu

Minimata

China USSR

N.Korea

S. Korea

Japan

Kyushu

born with the disease. Many of them had mental retardation, cerebral palsy, inability to speak, and other serious conditions as a result.

In 1968, a thirty-six-year-old former fisherman, Tsuginori Hamamoto, was stricken with the sickness. His parents had died of it in 1956. In 1973, determined to bring more attention to the problem, Hamamoto went to Stockholm, Sweden, where he explained the problems of Minimata to members of the United Nations Convention on Human Environment. Other families sued the Chisso company. In court, the company was found guilty and ordered to pay damages to every victim of methyl mercury poisoning in Minimata. Soon the factory closed its doors.

Other tragic cases of poisoning by methyl mercury and various toxic wastes have been discovered since that time. They show that the ocean and its bays cannot be taken for granted. The people of Minimata depended on the bay for food and work. The fertilizer plant depended on the same bay as a place for dumping wastes. As they painfully learned, its waters could not safely be used for both. This lesson was repeated, under different circumstances, in another part of the world in the 1970s.

The death of the Baltic Sea

This time, the problem surfaced in the Baltic Sea, an arm of the Atlantic Ocean linked to the North Sea. The Baltic has 163,000 square miles of mildly salty water and once was an important fishing source for cod, herring, and salmon. Surrounded by six heavily industrialized nations—Sweden, Finland, the Soviet Union, Poland, Germany, and Denmark—the Baltic has been subjected to tremendous discharges of wastes since trade began in the area in the fourteenth century.

In the 1950s, the Baltic's sparkling blue waters were described as a sailor's Garden of Eden. Fishing and recreational boats sailed up and down the coasts, and sunbathers flocked to the beaches. Seafood festivals and sailing regattas were frequent. In many areas, lush farms lined the shores between busy harbors such as Stockholm, Helsinki, Leningrad, and Copenhagen.

But the residents and visitors did not know that their playground was being turned into a garbage dump. After World War II, more than thirty thousand bombs and containers filled with mustard gas were tossed into the Baltic. Mustard gas, a poisonous chemical that can kill if inhaled, was a chemical warfare weapon used in both world wars. The Baltic also seemed like a good place to dump seven thousand tons of arsenic, a deadly chemical used in the manufacturing of agricultural pesticides. Cement containers holding this arsenic were tossed into the Baltic Sea in 1957.

The dumping did not rid the area of these dangerous chemicals, as the manufacturers and nations had assumed it would. Instead, the ocean currents moved the containers nearer the shoreline, and pressures of the deep water caused them to develop leaks. The poisonous mustard gas spilled from the containers and injured people who were fishing in the area. To make matters worse, some of these people brought up in their nets sixteen mustard gas bombs thought to have been placed forever on the sea's bottom.

This lethal catch meant that tons of fish had to be removed from sale on the food market. Those who made a living fishing saw their livelihoods slowly decline because they could not sell fish that contained the toxic gas and other poisonous chemicals.

In the early 1970s, after analyzing the water, marine biologists found that the centuries during which many nations had used the Baltic as a dumping ground had taken a great toll. The sea was low in oxygen to begin with because its water is exchanged very slowly with the water of the North Sea and eventually the Atlantic Ocean. This process occurs slowly because the two seas are connected by a narrow channel. Thus, it takes a long time before the stagnant, deoxygenated water is replaced with fresh, oxygen-rich water. Any wastes added to the sea accelerate the process of stagnation, or inability to sustain life.

Results of dumping

The dumping of untreated sewage, 80,000 tons per year of industrial wastes containing mercury and other heavy metals and chemicals, contributed to the death of the Baltic Sea. Other contributors were the runoff of poisonous agricultural wastes and the mustard gas. When any body of water dies, it no longer produces oxygen, and it cannot support plant or animal life. The Baltic Sea's ecosystem was thrown out of balance. As animal and plant species died out, others that depended on them also died out. The surviving fish stopped reproducing, and the ocean environment poisoned any plant or animal that tried to live in it. This process was gradual and not visible, so no one realized the Baltic Sea was dying until the mustard gas surfaced. The only marine, or undersea, life that remained was microscopic bacteria that needed no oxygen to live.

Similar processes are taking place in seas and coastal areas throughout the world. These situations have caused many people to question the actions of business and government and to worry about the ability of the oceans to survive the growing pollution by our society. The French pioneer of underwater exploration, Jacques-Yves Cousteau, wrote, "The very survival of the human species depends upon the maintenance of an ocean clean and alive, spreading around the world. The ocean is our planet's life belt." With the help of modern oceanography, scientists have identified many of the substances that threaten that life belt, and they are beginning to learn how to repair or at least lessen the damage.

2

The Manufacturing of Pollution

INDUSTRIAL WASTES, sometimes known as trade wastes, are the residues left over from manufacturing processes. Industrialized nations in Europe and developing countries in Latin America and the Far East create unknown quantities of industrial wastes. During the 1980s, U.S. industries created approximately 400 million tons of industrial wastes each year.

The U.S. Environmental Protection Agency (EPA) has estimated that almost one-fourth of this amount, or 176 billion pounds, is dangerous. By another estimate, more than 100 million tons of industrial wastes were dumped into the ocean each year by the United States in the 1970s.

Dangerous wastes

The danger created by industrial wastes is usually labeled as one of two types. These wastes may be hazardous, able to do harm by causing fires or explosions. Or they are toxic, chemically poisonous and able to cause serious sickness or death if they are ingested or breathed or some-

(opposite page) Manufacturing processes produce industrial waste. This waste disposal site in Washington state contains nuclear waste.

25

times even touched. Toxic elements are dangerous in any amount because living things have very low tolerances for them.

Since the Industrial Revolution of the 1800s, industries located on coastlines, estuaries, rivers, and bays in the United States have regularly dumped their wastes into coastal waters. Industrial waste disposed of in other ways also winds up in the oceans.

Inland factories may dispose of wastes in legal burial or dump sites called landfills, on vacant land, or by piping directly into rivers. Some liquid wastes are piped into sewage treatment plants. But the waste in landfills usually leaches, or leaks, into groundwater, the underground water that flows into rivers and streams that eventually join the ocean.

The ocean is the drain of all the water found on the surface of the globe and in the atmosphere. The dissolved or solid wastes dumped into rivers contribute new materials to the ocean.

Living things like birds, plants, and animals have a low tolerance for toxic elements. Any amount of a toxic element can be dangerous.

This dumping did not pollute the ocean ecosystems with large amounts of hazardous materials until societies became heavily industrialized.

Industrial waste enters the ocean in other ways too. For many years, ships have been able to obtain federal government permits to dump industrial waste directly into deep ocean areas, although a new law makes this practice illegal as of 1992. Ash produced by waste incineration also ends up in the ocean. This ash, sometimes containing toxic substances, is released into the atmosphere and falls back onto land or into the sea in the form of acid rain. Acid rain is natural precipitation full of deadly chemicals.

Significant amounts of industrial waste also enter coastal marshes, wetlands, and open seas by illegal "moonlight" dumping. Marshes and wetlands are land areas often near oceans that have their own unique ecologies involving many marine or aquatic plants and animals. Individuals, industries, and waste disposal companies dump their garbage secretly on land or at sea in order to save time and to avoid paying legal dumping fees.

Toxic waste kills life in inland waterways, too. A fisherman motors through a blanket of dead and dying fish in Escambia Bay, Florida. The fish were killed by industrial wastes, untreated sewage, and fertilizer runoff.

The wastes created from industrial processes require many years to decompose.

Because of this massive dumping of agricultural, municipal, and industrial wastes into the oceans, large amounts of toxic chemicals are changing the ecosystems of coastal areas and perhaps of deeper ocean areas. Natural by-products are recycled or eventually biodegrade, that is, they eventually decay or break down into their original natural substances, which are reabsorbed by nature. But the wastes of industrial processes do not biodegrade, or they take a long time, sometimes thousands of years, to do so.

Instead, toxins from industrial wastes may accumulate in the tissues of living organisms. Industrial pollution in the ocean has killed or contaminated beds of edible shellfish and marshes.

Contaminated fish are unsafe to eat. These poisons become more concentrated as they are transferred up the food chain, eventually to humans. In 1971, for instance, the U.S. Food and Drug Administration (FDA) had to destroy fifty thousand turkeys, eighty thousand chickens, and sixty thousand eggs that had been poisoned by chemical compounds called PCBs. All the poultry had been affected by PCBs in the fishmeal they ate. And in U.S. coastal areas near high concentrations of chemical waste, fish with tumors in their livers have been found.

Synthetic compounds

Most trade wastes result from new metal and chemical compounds created by industry. Hundreds of chemical factories produce the compounds used by other industries to manufacture dozens of products used daily in homes and businesses.

Industries have employed chemical elements since the early 1900s, but most of the compounds used today did not exist before World War II. At that time, chemists discovered and developed new compounds through synthesis, or mixture of two or more natural elements such as oil, chemicals, and metals. The result was many new, artificial chemicals and compounds not found in nature, known as synthetics.

Using synthetics, factories mass-produce thousands of common household items, including plastic products, clothing, medicines, cleaning agents, and paper items. Because popular demand for these products made them profitable for many industries, the number of products has increased dramatically each year. The amount of waste they create, both in being produced and in being used by consumers, has also increased. By 1990, the United States was actively producing fifty thou-

If mishandled, toxic substances can sometimes kill. For their own safety, these workers at a chemical-waste site wear protective clothing when handling toxic chemicals.

sand compounds and creating six thousand more each week.

The wastes that result from the manufacturing of synthetics or from using them to make consumer products can be toxic. If improperly handled, some compounds can poison people and cause serious illness or even death. Some have been identified as carcinogenic, causing cancer in animals or humans. If improperly stored, hazardous compounds can explode, burst into flame, or corrode, that is, eat through their containers or other substances. Hundreds of these dangerous compounds are found in industrial wastes that are dumped in the ocean. The two most dangerous sources of industrial ocean pollution are metal compounds and chlorinated organics, that is, chemical compounds.

Metal compounds

Many metals, such as mercury, copper, lead, zinc, and aluminum, occur in nature. They are safe in balanced amounts in a particular environment as long as they are bonded, or stuck, to the soil. But in synthetic metal compounds, this bonding can break down, and the metals are released to be captured in the tissues or muscles of animals, where they do great harm.

That is what happened in Minimata, Japan, when the manufacturing process converted the mercury waste to methyl mercury, which poisoned many of the animals and people who ate the fish from the water polluted by this waste. Hazardous and toxic compounds such as methyl mercury are a result of synergism. Synergism is a chemical process in which two compounds act simultaneously to produce effects far greater than would be caused by either alone.

Lead, mercury, cadmium, chromium, copper, selenium, and silver are known as heavy metals because they are at least five times heavier than water. These heavy metals can enter the human body when a person eats food, breathes fumes, or touches objects that contain them. Compounds of heavy metals are the most dangerous to plants, animals, and humans. Heavy metals in high levels have been shown to retard growth and increase the death rate of marine animals including shrimps and worms.

Copper poisoning

Copper is one heavy metal that has harmed ocean ecosystems. In March 1965, industrial copper sulfate was poured into the North Sea along the Dutch coast, increasing the natural concentration of copper in the seawater more than a hundred times. After this waste traveled in a solid mass along the coast, 100,000 dead fish washed ashore, whole mussel beds died, and many species of plankton disappeared. In other areas, at other times, oysters have been made green and inedible by the copper in the water.

The most dangerous synthetic chemical compounds are those containing the chemical chlorine combined with petroleum. These substances

greatly increase in concentration with each step up the food chain. Many scientists today are concerned about chlorinated compounds that contain cancer-causing substances, particularly pesticides used by farmers to kill insects that harm their crops.

One reason pesticide chemicals are so dangerous is that they have been designed to kill. They are developed to easily enter the body of an insect and kill it from outside or inside. People use pesticides without considering that they work the same deadly way on other living creatures, including themselves.

Deadly DDT

It has been illegal to use DDT (dichloro-diphenyl-trichloroethane), probably the most famous chlorinated compound pesticide, in the United States since 1972. However, its effects will be with the world's ecosystems for many years to come. In the years from 1944 to 1968, about 1.2 million metric tons of DDT were manufactured in the United States and sprayed on agricultural fields throughout the country from the ground or the air. As much as a quarter of the annual production of DDT compounds was transported to the world's oceans by precipitation, even though it was initially sprayed over land.

Residues of DDT have been found in fish-eating birds such as pelicans and cormorants in Great Britain and the United States. More surprisingly, DDT remains have appeared in animals and fish that live where the pesticide is not known to have been used: in Antarctic penguins, Arctic krill, Atlantic phytoplankton, and North Sea cod. The growth, reproduction, and death rate of all these organisms were adversely affected by DDT.

DDT is so dangerous because it takes decades

Some of the chemicals sprayed on trees and plants eventually find their way to the oceans.

to biodegrade. It has been shown to inhibit photosynthesis in phytoplankton and therefore to put all other marine life in jeopardy. Scientists became really concerned about DDT when they found evidence of it in human breast milk. Other researchers have linked this pesticide to possible disorders of the human nervous system and the liver and even to cancer. DDT is still manufactured in the United States and sold to other countries to combat insects such as the malaria mosquito. As a result, it is still being added to ocean ecosystems throughout the world.

The perils of PCBs

Another group of chlorine-petroleum compounds is polychlorinated biphenyls, or PCBs. Discovered in 1929, 80 PCB combinations are used extensively and 210 are possible. The combined chemicals make PCBs highly resistant to heat so they form a good base for insulation in industrial wiring, air conditioners, transformers in power plants, and fluorescent light fixtures.

PCBs are not easily dissolved in water, and they have high chemical stability, meaning they do not break down quickly. Both these qualities make PCBs desirable in industry, but they are exactly the qualities that make them harmful pollutants. They make PCBs extremely slow to biodegrade.

In 1966, a Swedish scientist, Soren Jensen, was researching the pollution problems of Lake Erie in the United States. He linked PCBs to fish-eating birds and found that PCBs were passed on to as many as six generations. Baby birds that had inherited this poison were born with deformities such as twisted beaks, clubfeet, twisted spines, and brains and hearts developed outside the body.

Since then, PCBs have been detected in many species of fish. In Great Britain they have been identified in the corpses of sea lions, aborted sea lion pups, and in abnormally thin shells of seabird eggs. Although production of PCBs has been halted throughout the world since 1979, materials containing PCBs are still in use, and PCBs from past waste discharges are still entrenched in plankton, seaweed, flying fish, dolphins, sharks, Arctic and Antarctic mammals, and ocean sediments.

PCBs in New Bedford

Some scientists say that the residues of PCBs will be present for at least five hundred years. The International Agency for Research on Cancer, part of the World Health Organization, says PCBs are carcinogenic to humans. The varied effects of PCBs have been felt first-hand by at least one fishing community in the United States, New Bedford, Massachusetts.

New Bedford, once the richest fishing town on the East Coast, has been a fishing center for several centuries. It is the city from which author Herman Melville launched the fictitious whaler ship *Pequod* in his famous novel, *Moby Dick*.

In the 1930s, two electrical manufacturing businesses moved into New Bedford, and both made capacitors, an electronic part that uses PCBs. In the 1970s, the companies used nearly two million pounds of PCBs per year. For all those years, they dumped waste into the Acushnet River and the estuary of Buzzards Bay. As a result, PCBs soon saturated the sediment, the soil beneath the water, of the 985-acre harbor, reaching 100,000 parts per million (ppm).

It was not until 1974, when scientists were measuring the amount of oil in the harbor, that they accidentally discovered this high proportion of PCBs. By the early 1980s, thousands of acres in the bay and in the adjoining Atlantic Ocean had to be closed to harvesting shellfish, finfish, and lobsters. This seriously damaged New Bedford's economy, which is built on fishing and on attracting tourists to its beaches.

Although the United States government banned manufacture of PCBs in 1977 and the two electrical manufacturers on the Acushnet stopped using them shortly after that, PCB levels in the harbor did not drop. The compound apparently had attached itself to sediments in the sewers, and it still continues to move into ocean waters. People in the area have been measured and found to have extremely high PCB levels in their blood.

In dozens of studies, the federal and state governments have measured PCB contamination in New Bedford and tried to find ways to erase the damage. Other fisheries in New England, New York, and California have been closed because of this toxic contamination. Thus, even banning a known toxic compound from ocean dumping does not put an end to its power to do damage. The long-lasting effects of toxic pollution are es-

Industrial dumping has severely damaged the water and sediment of the harbor in New Bedford, Massachusetts. Dumping there began in the 1930s.

pecially serious in regard to radioactive materials.

Scientists have learned to harness atomic power to produce energy and military weapons, but they have not yet solved the problem of safely disposing of the resulting wastes. These wastes are often placed in containers of steel, lead, or concrete, which are then dumped in the ocean or buried deep in the earth. But the radioactive wastes continue to boil with their own energy and remain dangerous for long periods. They may penetrate the containers that are supposed to hold them. They can also penetrate human and animal tissue.

The deadly and long-term harmful effects of this atomic radiation were seen in the victims of the atomic bombing of Hiroshima and Nagasaki in World War II and in accidents that have occurred in plants producing nuclear energy and weapons. Scientists have not determined exactly how much radiation humans can be exposed to safely. Some think any amount is dangerous.

High- and low-level wastes

Radioactive wastes, sometimes called "hot garbage" or radwastes, are divided into two categories. High-level wastes remain active and dangerous for hundreds or even thousands of years. Low-level wastes become harmless in a shorter time, from a few seconds to a few years. The most toxic radioactive wastes are plutonium 239 and uranium 235. Others that have long-lasting effects are strontium 90, iodine 131, and cesium 137.

All of these high-level radwastes are bioaccumulative; they remain a danger to life for thousands of generations. Scientists believe that plutonium 239 will remain toxic for up to half a million years. Its toxicity is so powerful that one millionth of one gram inhaled is enough to cause cancer in a human being. One thousandth of a gram inhaled will cause death within a few hours. Even after death, the victim's body remains radioactive and dangerous. People killed by massive doses of radiation in nuclear power plant accidents are buried in lead coffins for the safety of others.

Doses of high-level radwastes that are not large enough to kill animals and people can still cause severe burns or miscarriages. Even minute doses can cause loss of hair, radiation sickness, cancer of body organs, or leukemia, a form of cancer that affects the blood cells and the bone marrow. Other effects include injuries to the bones and the thyroid gland, which regulates growth and the body cells' use of energy.

One effect of radwastes and other toxic chemicals is that they may be mutagenic, meaning they may cause changes or mutations in the body's

Drums of nuclear waste washed ashore. Even minute doses of such wastes can be harmful or lethal to people and animals.

DNA. DNA is the human hereditary material. Changes in DNA may result in deformed or retarded babies or in physical disorders or diseases passed on through several generations.

How radwastes reach the ocean

Nearly all nuclear reactors or power plants are located on coasts or near large bodies of water. These plants throughout the world constantly release small quantities of radioactivity to the aquatic environment. From 1946 until 1983, these plants also used the oceans to dispose of stockpiles of accumulated wastes.

In the United States, nuclear reactor wastes, including contaminated laboratory equipment, clothing of workers, and sweepings from floors, were encased in cement, then packed into steel drums. Thousands of these containers were dumped off the East and West coasts and in unknown deep ocean areas. But ocean currents and earthquake shock waves carried many drums to

Most nuclear reactors and power plants are located near large bodies of water. For this reason they can pose a significant threat to the world's oceans.

other areas, and some have corroded and leaked.

From research on the atomic bomb tests conducted in the waters of the Pacific Ocean in the mid-1940s, scientists learned that radioactive substances released into the ocean melt into ocean sediment and enter the food chain. For example, strontium 90, used in early atomic weapons, has been found in the food chain of the South Pacific. Dr. Jerold M. Lowenstein, past chairman of nuclear medicine at the Pacific Medical Center in Seattle, Washington, expects traces of strontium 90 to remain in the food chain until at least the year 2026.

Bomb tests

About two thousand miles southwest of Hawaii are the Marshall Islands, made up of about thirty atolls, each comprising dozens of tiny coral islets in lagoons. This was once a tropical paradise, where people lived off the fish they caught in clear, clean waters. Between 1948 and 1958, the United States exploded atomic and hydrogen bombs on these islands to test them. In 1957, the 166 people living on Bikini Atoll were removed before twenty explosions were made.

After the bomb tests, people living on the Rongelap Atoll, one hundred miles away, experienced radiation sickness. The eighty-six residents were moved away. Three years later the U.S. Atomic Energy Commission (AEC) declared Rongelap safe for rehabitation and moved the people back.

But during the next year, the miscarriage rate of exposed Rongelap women was twice that of unexposed women. By 1969, the average radiation levels of the Rongelap people were still ten times those of the people on a noncontaminated island. Thyroid problems appeared in almost every resident who had been less than twelve years old in 1954. Growth retardation, thyroid malignancies,

These dead fish washed up in an area close to New Jersey's first nuclear power plant in Lacey Township.

The first atomic blast at Bikini Atoll sends a mushroom cloud high into the air. People living nearby experienced radiation sickness from the blast.

and cancers including leukemia also emerged.

By 1968, Bikini's former inhabitants wanted to go back to their island. The AEC promised that the island was safe for their return and that "there is virtually no radiation left." Nonetheless, the AEC insisted that the houses on the island be built with six-inch-thick cement floors to protect against radiation. Some contractors, trying to save money, used sand from nearby contaminated beaches in these floors, further exposing the people rather than protecting them.

The residents of the atolls had traditionally fished year-round, and their land and tropical climate had been lush and fertile. But now all food and water for consumption had to be imported. They could not eat the crabs and fish in their waters.

Throughout the 1960s and 1970s, health prob-

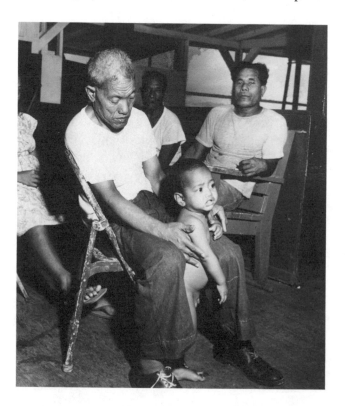

Radioactive ash showered these natives of Rongelap Atoll, one hundred miles from Bikini Atoll.

lems believed to be related to radiation appeared among residents of many of the atolls where atomic tests had been conducted. The people of the Marshall Islands accused the AEC of downplaying and even hiding the potential dangers of the radiation exposure. The agency was monitoring human health and measuring radiation levels of the animals, fish, and plants. But residents charged the AEC with not being truthful about what it had found and of treating them like animals in a scientific experiment.

In 1977, more than twenty years after the bomb tests, a study found that all living creatures on Bikini exceeded U.S. guidelines for safe radiation levels. Then the U.S. Department of Energy (DOE) put Bikini off-limits for the next sixty years. In 1979, the AEC told the Rongelap people that their northern islands were too radioactive, though they had been using them for twenty years. The people were advised to evacuate. No one knows when the area will be safe enough to support life again.

3

Everyday Ocean Pollution

ALTHOUGH INDUSTRIAL wastes in the ocean pose serious threats to the ecosystems and to human health, wastes generated by the daily activities of human beings are the largest source of ocean pollution. As the world population increases, possibly up to seven billion by the year 2000, the amount of wastes people produce in their daily activities is also growing to serious proportions. This is especially true in the industrialized nations, which have more waste because of their modern products and standards of living. Although the developing countries have more people, they produce much less waste because they have far fewer goods to dispose of.

Wastes created in everyday living

People today create a steady stream of wastes in their daily lives. These include natural body wastes, or raw sewage, and the garbage or trash from homes and businesses. Another category of human waste is the medical supplies, materials, and human tissues contaminated by germ-carrying bacteria from people being treated in hospitals, doctors' offices, clinics, and at home. Garbage includes the nonbiodegradable products

(opposite page) Human activity generates the largest source of ocean pollution.

43

of modern technology, primarily plastics of all types, used in homes and businesses. As with other types of wastes, for many centuries people have dumped these materials into the oceans trying to put this trash out of sight and out of mind.

The oceans' coastal areas have been seriously affected by pollution from human wastes. Because so many people live near shorelines, these coastal waters have shown much damage in the last forty years. Today more than 40 percent of the U.S. population lives within fifty miles of a saltwater shoreline.

As a result, a cycle of toxic pollution is created. More disease-carrying human waste is being expelled into the ocean near these human-inhabited areas, and the damage to shoreline ecosystems is accelerating. Because this pollution exists near areas where many people live, more people are likely to be affected by it. And, through their own natural bodily functions, these people then contribute larger amounts of potentially hazardous waste.

Sewage wastes

The meaning of the word sewage has changed over time. Once it referred only to bodily wastes excreted by humans. During the Victorian era in the late 1800s, the word sewage offended people of culture, so they called human wastes effluents. Today effluents can be any kind of waste product. Sewage refers to all solid and liquid wastes that flow into sewer drains after being flushed or poured down toilets and sinks. After being treated and cleaned, this sewage is sent into natural waterways, which lead to the oceans. But the treatment process does not remove all toxins.

In the United States, the wastes piped into public sewage systems from homes, businesses, and other buildings used by people include many

"I guess they gave up trying to clean up all the medical waste on this beach."

types of toxic pollutants. For instance, sewage systems are the receptacles for soapy water from washing machines and showers and for food particles from garbage disposals. People also pour used cleaning liquids down kitchen and bathroom drains.

In an average city of 100,000 people, it is estimated that three tons of toilet bowl cleaners and more than thirteen tons of toxic household cleaners are piped from homes each month. Toxic wastes also come from the products discarded into the water system by more than 130,000 commercial businesses nationwide, such as laundries, beauty salons, printing establishments, car washes, food manufacturers, and markets.

Problems of aging sewage systems

The most serious pollution from sewage in the United States is found in the Northeast, where human wastes have been dumped into bays and other ocean waters since colonial times and where people live closely together in large cities. Another complication in these older communities is that the growing volume of wastes is overtaxing their aging sewer systems. Sewer systems designed in the early 1800s were created to dispose of street filth by letting rain wash it into drains. Later, pipes were added to carry human wastes from flush toilets. These systems simply spilled their liquid and solid wastes into the nearest available streams, lakes, rivers, or bays, and some still do so today when it rains heavily.

In most places, new pipes have been connected with the old systems to carry raw sewage wastes to treatment plants. Each year, these systems carry 300,000 tons of liquid and solid residues that must be disposed of, and the amounts are increasing.

In countries where raw sewage is not carefully monitored, disease-carrying bacteria and virus germs may get into the water that people drink or bathe in, which transmits them through entire populations. Disease bacteria can also survive for as long as five years in shallow ocean water or by clinging to rocks or bottom sediment.

The amount of liquid and solid residues treated at sewage-treatment plants increases every year.

Pollution has caused many oceans, bays, and other bodies of water to be closed to swimming and other recreation.

These disease-bearing bacteria or viruses can be stored in the tissues of shellfish and then infect other animals or people who eat them. Americans are still stricken every year with hepatitis and viral gastroenteritis, a severe stomach disorder, from shellfish contamination. Some even die. However, most U.S. communities put sewage through a treatment process that eliminates most bacteria.

Sewage treatment plants

Treatment of sewage wastes involves a two-step cleaning process. The first, or primary, process separates the solids from the liquid wastes by gravity, and the liquid may be piped directly into bodies of water for disposal even though it still contains toxins and bacteria. In secondary treatment, colonies of living microscopic bacterial organisms are placed into a tank with the primary-treated wastes. Just as they do in nature, these bacteria decompose or break down the remaining floating waste. Oxygen is then added to speed this process. After primary and secondary treatment of sewage, what remains is sludge, the semisolid residues of human waste, which are filled with disease-carrying germs. This sludge is often discharged into a nearby body of water.

By 1988, ten trillion gallons of treated sludge were being discharged into marine waters through 578 of the 15,500 U.S. treatment plants each year. On top of that amount, sewage reaches the oceans from rivers, and raw sewage is also dumped into marine waters. A 1988 U.S. law required that sewage treatment processes be upgraded so the sludge would be less dangerous.

Sludge disposal

Today most European countries export or incinerate their sludge wastes. Specially designed ships burn wastes without creating polluted residues in the oceans or the atmosphere.

In other areas, societies have figured out ways to reuse secondary-treated sludge. Throughout history, people have collected human wastes to use for fertilizer. Today China treats this waste and deposits it back on farmland. In the United States, the city of Cleveland, Ohio, dehydrates, or dries, its sludge, incinerates it, then spreads the ash residue on forest lands to aid tree growth. Milwaukee, Wisconsin, sells dried sludge for fertilizer. Other states clean it and use it for ground fill or to cover abandoned landfills.

But Great Britain and the United States still permit dumping of sludge wastes into the oceans. Although both countries are phasing out this process, large coastal cities, including New York, Boston, and Los Angeles, still pipe sludge into their bays. Los Angeles County in Southern California spills more than 107,000 tons of sludge

each year directly into the Pacific Ocean through an outfall, the open discharge end of a long pipeline connected to several treatment plants.

By 1977, fifteen sewage treatment plants along the 127-mile coastline of New Jersey were discharging 130 million gallons of sludge per day directly to the Atlantic Ocean. At the same time, New York City piped twenty million gallons of treated sewage and more than half a million gallons of partially treated sewage through outfalls into its harbor each day. The water from there flows directly south and joins New Jersey waters.

New York also sent barges and ships loaded with tons of garbage and sludge daily to be dumped at a bight, a legal dumping site. The bight was twelve miles offshore for many years and was

A tug boat hauls a barge loaded with tons of Long Island garbage. Four states and two countries barred the barge from entering their ports.

The garbage on this fouled coastline includes broken-up furniture and other household items.

then moved 106 miles out to sea. In the 1980s, a series of incidents brought international attention to the pollution of this coastline.

In the summer of 1987, beach after beach in New Jersey was closed as incoming tides washed in tons of discarded wastes from the Atlantic Ocean. This foul-smelling, unsightly catch included sludge from New York and New Jersey, millions of dead, decaying fish and dolphins, garbage, and medical wastes. The sludge contained high counts of coliform, raw sewage bacteria. The dolphins showed traces of PCBs and heavy metals. The garbage included decomposed rats and plastic throwaway products. The medical wastes consisted of bloody bandages, vials of blood, animal tissue, and used hypodermic needles. Tests of water revealed high bacteria counts.

The following summer, this sickening harvest reached the beaches of New York state. Some beaches were littered with balls of grease and sewage two inches thick. Some of the blood in the vials tested positive for the AIDS and the hepatitis viruses, although investigators do not believe these vials could have infected anyone. An investigation of two hundred square miles of ocean bottom revealed nothing but black, mayonnaiselike sediment.

Similar debris has washed up elsewhere in the country and the world. Seventeen states have reported dangerous medical wastes appearing on their coasts. The public outcry over this pollution culminated in passage of the Ocean Dumping Act of 1988, which regulates disposal of sludge and industrial wastes.

Lost business

This repulsive harvest had a disastrous effect on New Jersey's tourist economy. Lost business was estimated to be in the billions of dollars. Near the New York bight, diseases found in shellfish were traced to dumping of sludge. An unusually high rate of dolphin deaths was also related to the contaminants from garbage and sludge.

Investigations could not pinpoint a single source of this horror. It can probably be attributed to longtime disposal practices of all these types of human-generated waste in communities along the New York-New Jersey coast. New York state's hospitals, clinics, and medical labs generate 125 tons of infectious wastes each day. Authorities searching for the source of this medical debris found that it was the result of moonlight dumping by waste disposal companies. These businesses were entrusted to manage medical wastes without endangering human life or the environment.

Medical wastes

The New Jersey shore disaster brought public attention to the serious problem of medical waste polluting coastal waters, both with unsightly garbage and with potentially dangerous disease bacteria. The EPA estimates that hospitals across the country generate around three million tons of infectious wastes each year. This figure does not

These syringes and blood vials were found along New York state beaches in the summer of 1988.

include the medically contaminated materials discarded by laboratories, doctors' and dentists' offices, or thousands of small clinics and medical services. Much of this is incinerated, but some winds up in the ocean.

The AIDS epidemic and the awareness of increased drug use have brought an outcry for federal regulation of the disposal of medical waste and needles. New regulations are being considered by the government.

While the large volume and the potential danger of medical waste is a relatively new dilemma, garbage has always been with us. And much of it winds up in the oceans, so long treated as the world's trash can.

Garbage

The United States produces 180 million tons of garbage per year. If this amount were loaded on ten-ton trucks lined bumper to bumper, they would circle the globe almost three times. According to researchers, the average American throws away three and a half pounds of garbage or trash every day. This garbage includes clothing, furniture, containers, packaging materials, appliances, magazines and books, diapers, and toys. These items are made of every conceivable material, much of it nonbiodegradable synthetic compounds.

Cities except New York stopped dumping garbage such as this in the oceans in 1934 mostly because it looked so bad. Most of that garbage now goes into legally authorized landfills. There the trash is buried, incinerated, or recycled. Some states transport their trash, or municipal solid wastes, to landfills in other states or by ship to other countries.

Until new laws took effect in the 1980s, however, tons of garbage were dumped daily from military vessels, cruise liners, oil tankers, fishing fleets, and offshore drilling sites. Metals, glass, and some other items would sink out of sight, though they still affected the undersea communities.

Plastic—killer of ocean life

Plastic is another story. The plastics industry in the United States employs more than one million workers who produce an output nearly double the combined total of steel, aluminum, and copper products. The variety and usefulness of these thousands of plastic products are a result of industry's development of synthetic compounds. These items are designed to be durable, and it is that very durability that has made them an environmental problem. Plastics are buoyant and continue to float and drift, sometimes for years, with ocean currents. Thus, the oceans have carried plastic garbage all around the world. American plastic-coated milk cartons have been found on Norwegian beaches, and French mineral water bottles on American beaches. By 1985, it was es-

A park supervisor cleans a stretch of sand at Rockaway Beach, New York.

timated that each day, commercial vessels alone dumped 450,000 plastic containers and other items into international waters.

Recreational swimmers and sunbathers may sometimes leave their trash at the beach, but they are not accountable for the bulk of plastic items that litter beaches. Most of it is washed up from the sea, where it has been dumped as garbage. The shorelines that accumulate the most plastic are those with the most shipping activity. In 1987, one three-hour search by volunteers over a 157-mile stretch of beach in Texas produced 115,727 separate plastic items, far more than could have been left by beach visitors. These items included plastic bags, drink bottles, six-pack rings, lids, disposable diapers, milk containers, and tampon applicators. The dilemma of plastics pollution proves that there is no such thing as a throwaway society. What is thrown away comes back to make the world less attractive and more dangerous.

The real victims of plastic

Plastic trash on the beaches looks ugly, annoys and distracts bathers, and is costly to clean up. But it actually poses no direct threat to human health and life. However, plastic is a nightmare for marine animals. It kills thousands of fish, seabirds, turtles, otters, dolphins, porpoises, penguins, and whales every year.

It inflicts this horror in simple ways. The animals eat or swallow plastic items, mistaking them for food, and they starve or develop deadly infections. Or they become entangled in floating plastic, which often results in drowning, suffocation, or a slow death from starvation or exhaustion.

Durable plastics cannot be digested or expelled by living creatures. This material blocks their intestines, causing infections and eventually death. Seabirds diving for food may mistake

Plastic debris and waste materials litter oceans and inland bodies of water, too. Here a volunteer records the kinds of plastic found at Chicago's Montrose Beach on Lake Michigan.

A turtle entangled in a fishing net. The turtle's inability to move will cause it to suffocate and die.

floating particles of plastic for food, or turtles may see plastic bags or gloves as jellyfish. In 1984, a stranded baby whale was examined after it had died. It had an infection in its abdominal cavity caused by a thirty-gallon garbage can liner, a plastic bread wrapper, and a corn chip bag lodged in its stomach.

Floating fishnets, webbing, packing straps, ropes, six-pack holders, fishing lines, and motor gaskets ensnare marine mammals, birds, and fish. Nets hook on the gills of fish, eventually cutting off oxygen and suffocating them.

Species have actually shown declines in significant numbers as a result of plastics. Besides causing animals to suffer, their untimely deaths in this fashion will change the ecosystems. This will ultimately affect all living organisms including human beings.

Death by plastic

In 1976, the seal population was decreasing by 4 to 6 percent a year in the Pribilof Islands in the Bering Sea west of Alaska. Scientists from the

This large, healthy seal was killed by a narrow, thin piece of plastic.

National Marine Mammal Laboratory concluded that as many as forty thousand seals were killed every year by plastic in the ocean environment.

The naturally curious animals saw the plastic debris as playthings. They became entangled in plastic netting or packing straps, their movements were restricted, and they could not feed normally. Unable to move, they drowned, starved to death, died of exhaustion, or died from infections from deep wounds caused when the plastic material tightened around their backs and necks. Countless seal pups placed their heads through the rings of six-pack holders or fishnets and grew into these plastic collars, which severed their arteries as they became larger.

Whales starve or drown after being entangled in hundreds of feet of abandoned or lost commercial fishing nets. The victims of death by plastic include great numbers of endangered species such as the brown pelican, puffins, sea lions, and rare turtles. Tony Amos, an oceanographer at the University of Texas Marine Science Institute in Port Aransas, has patrolled a seven-and-one-half-mile stretch of beach along Mustang Island in the Gulf of Mexico for at least twelve years. In 1988, he examined 110 loggerhead sea turtles that had washed ashore and found that 57 percent of them had swallowed plastics. On the beaches near Corpus Christi, he has found garbage originating from sixty countries, as far away as Finland and the Philippines. Amos estimates that up to 90 percent of the plastic items he finds are dumped off ships.

Amos tells of a Minke whale in the Gulf of Mexico that came ashore on the Matagorda Peninsula, still alive, but "completely emaciated." It died shortly, and scientists found that it had nothing in its two stomachs except a piece of plastic that had lodged in the passageway between them. Less than one ounce of plastic had blocked the passageway and caused the four-ton animal to starve.

An international law that went into effect in December 1988 has reduced this careless disposal of plastic and led to cleaner beaches. And in 1991, Maine outlawed the sale of containers bound together with plastic devices. But this has not stopped the harm done by plastics. These substances do not biodegrade or sink. Even as dumping declines, the plastics thrown into the ocean in the past will continue to float around the world, ensnaring unsuspecting animals wherever they go.

4

When Oil and Water Mix

OIL IS THE primary source of energy in the world. Oil, or petroleum, was created as the remains of marine plants and animals gradually turned to oil in the sedimentary rocks deep in the earth millions of years ago. Oil provides the gasoline and other fuels that run vehicles, ships, and aircraft. It heats homes and operates industry. Great quantities of petroleum are needed in manufacturing cosmetics, medicines, plastics, paper products, and numerous other items. The extracting, transport, buying, and selling of oil make up one of the largest businesses in the world. Many countries in the Middle East, South America, and other areas depend on an oil-based economy.

Oil travels the world

Because it is in demand in all areas, huge amounts of oil are imported and exported around the world. The oceans are tied closely to the oil business because oil may be extracted from deposits in the ocean, and it is shipped from one country to another on more than eight hundred oil tanker ships. These ships, some carrying as much as 200,000 tons of oil, must pass through narrow and dangerous straits and travel up and down

(opposite page) Since most oil transport occurs on the sea, the world's oceans are especially susceptible to oil spills and resulting environmental damage.

Much of the world's oil is extracted from the sea by use of oil platforms. About eight hundred platforms were being used worldwide in 1990.

coastlines into city waterways and bays. This offers many opportunities for oil to be spilled and to pollute ocean waters. Oil is also transported by pipelines, which can carry it for many miles to waiting tankers or storage tanks. The United States alone maintains almost 200,000 miles of oil pipelines.

Oil companies are constantly searching for new sources of oil. More than half the world's known oil reserves are located in underground deposits found in the dried-up, sandy seabeds of the Middle East. Geologists are aware that large deposits also lie beneath the ocean floor.

Leaks and oil drilling

Many companies search for oil in the waters near seacoasts, drilling along coastlines. In the ocean, oil extraction is accomplished by using oil platforms, or rigs. These large metal structures are built close to shore and anchored to the ocean bottom. About eight hundred offshore oil rigs were in operation around the world in 1990.

With such large amounts of oil constantly moving through the oceans, it is not surprising that spills occur. Even before oil drilling began, natural seepages of oil occurred, producing great pools of tar or asphalt in bodies of water. That process formed the La Brea Tar Pits in California, and the Dead Sea in Israel contains a seepage that was identified in biblical times.

Natural events also can cause oil to spill. Earthquakes can rupture pipelines. Hurricanes, storms, and fogs can cause barges and tankers to travel off course and damage their hulls, leading to oil spills. Many oil experts believe that the amount of oil spilled by nature each year is far greater than the amount spilled by people.

In addition, tankers routinely clean out their holds in the ocean, releasing oil into the water.

For example, routine U.S. oil and gas production dumps 1.5 million gallons of oily water into the Gulf of Mexico each day. Fire, explosions, oil rig blowouts, and leakages from pipes and storage tanks are also causes of spills.

One of the biggest blowouts happened when the Ixtoc well in the Bay of Campeche, Mexico, exploded and caught fire. It burned continuously from June 1979 until March 1980 before it could be capped. While burning, it gushed 600,000 tons of oil into the sea, and ocean currents carried some of this oil as far as the coast of Texas, hundreds of miles away, destroying marine life and wildlife all along the way.

Whether the oil is spilled by human error or accident of nature, large spills damage beaches and poison millions of animals. The largest and most serious oil spills are the result of human error. In the late 1980s, four major oil spills occurred off U.S. coasts, including eleven million gallons of crude oil covering eight hundred miles of Alaska shoreline.

Accidental oil spills

When tanker captains misjudge their courses, their ships sometimes ram reefs or rocks that can rip the hulls of the ships. When this happens, tons of petroleum pour out in one spot. In a few hours, an ecosystem and sometimes a human community are changed forever.

One of the most publicized tanker accidents occurred on March 8, 1967, when the giant supertanker *Torrey Canyon* passed near Land's End on the coast of England. Although it was a clear day, the captain miscalculated his course and ran into Seven Stones Reef. The hull was sliced open, and 119,000 tons of petroleum poured into the water. When this gooey mess washed up on British beaches, all the oysters and other shellfish in the

The hull of the supertanker Torrey Canyon *was sliced open off the coast of England, spilling 119,000 tons of gooey petroleum.*

area died, along with thirty thousand birds.

Between 1967 and 1989, the fifteen largest oil spills in the world spewed more than fourteen million barrels of oil onto the seas and the land. Although human beings are seldom killed in an oil spill, these disasters permanently damage ocean environments, kill millions of animals, and lead to loss of people's property and livelihoods.

In 1988, there was only one area of U.S. waters that had never been touched by oil. This was the pristine waters of Prince William Sound, Alaska. The trans-Alaska pipeline is one of the most important means of transporting U.S. oil. The petroleum extracted in northern Alaska is piped miles away to enormous supertankers in the bays of the northernmost state. Soon Prince William Sound would become known as the location of an enormous ecological tragedy.

The spill that could not happen

Valdez Harbor in Prince William Sound opened in 1988. The local people argued against opening the harbor. They feared the potential harm that an oil spill could cause to their clean and fragile region. The cold waters of the sound are home to concentrated populations of whales, porpoises, seals, sea lions, and ten thousand otters. Four hundred thousand resident birds are joined each spring by millions more migrating seabirds, water fowl, and shorebirds. Brown bears, deer, and bald eagles live along the forested shores. The sound, or bay, was then a prime area for salmon, herring, shrimp, and bottom fish, and the fishing industry expected a harvest worth $120 million in 1989.

But the oil companies said they needed this harbor. It is the only U.S. seaport able to receive and berth today's enormous supertankers, which sometimes reach sixty feet below the water's surface. These great ships are also too long to turn and maneuver in a normal-sized bay. In smaller harbors, they must anchor in the deep ocean while being unloaded. Then the oil is pumped from the ships through pipelines to storage bins

Prince William Sound was a clean, natural region before the Exxon Valdez *oil spill.*

In a matter of hours, oil from the Exxon Valdez *polluted more than twelve hundred miles of Alaskan coastline. The spill killed thousands of birds and animals.*

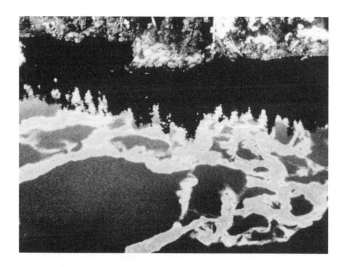

on land. Small spills are not uncommon from breaks in the pipelines.

The oil companies told the residents of Prince William Sound that no major oil spill could take place there. The most careful safety precautions were in effect. And if a spill should happen, Exxon Corporation, the world's biggest oil company, assured the Alaskans that it had equipment standing by to stop an oil flow immediately.

The toll taken by an oil spill

Shortly past midnight on March 24, 1989, the fully loaded *Exxon Valdez* oil tanker slowly headed into Prince William Sound. The tanker hung five stories beneath the surface of the water, and if it followed its assigned path, it would be safe from the jagged boulders on the ocean floor. But the ship went off course. By the time the crew realized it was headed for the rocky Bligh Reef, there was not enough time to stop it.

As soon as the hull scraped the reef, oil began leaking out of huge gashes in the tanker's hold. About 42,500 tons of crude oil oozed out of the tanker, fouling more then twelve hundred miles of Alaska shoreline in just a few hours. It was the largest amount of oil ever spilled in U.S. waters. The black, slick material lay on the ocean like a filthy blanket and was spread farther by the winds.

Animal victims

"Never before has such a sensitive natural showcase been so massively and suddenly violated," wrote one observer. The toll of damaged wildlife is stunning. Thousands of animals died on the first day. One local resident counted 650 dead birds on one beach, and thousands more birds weighted down by oil drowned and sank to the bottom of the sound. Oil-soaked bird corpses washed up on beaches 125 miles away. An estimated 580,000 birds died. Animals that ate the dead birds also ingested oil and died.

Bodies of some 109 Alaskan bald eagles were found on the beaches, and many more certainly drowned. Only a few thousand of these eagles

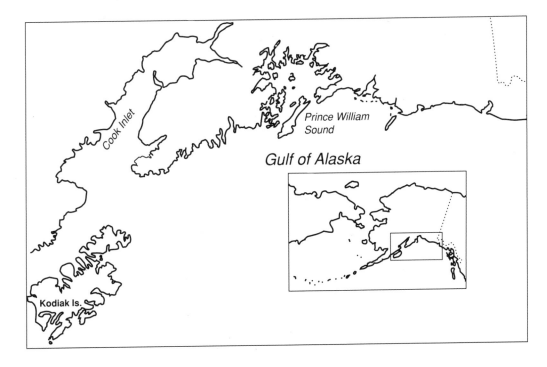

Three days passed before cleanup crews began a full-scale mop-up along the sound. Here a worker sprays hot water to remove the oil from a cobbled shore.

were known to exist before the spill. The reproductive cycles of the survivors were damaged, and scientists warn that the spill may have doomed the species. Pilot whales, too, suffered great losses, and many survivors miscarried in 1989 and 1990.

Sea otters were the hardest hit. The National Oceanographic and Atmospheric Administration (NOAA) estimates that as much as half the otter population in the sound may have died. Many of these animals drowned, others were poisoned by oil or killed by hypothermia, or freezing, because their oil-soaked fur could no longer insulate them from the cold waters.

Deer, bears, and other mammals on the nearby land suffered after eating seaweed and beach plants or oil-soaked animal carcasses on the beaches. Oil concentrations in shellfish will eventually be passed along to people. Some species of kelp were destroyed, imperiling the food chain.

Recreational use of the beaches was halted for a while. Herring fisheries had to be closed, and

millions of baby salmon were killed. NOAA estimated economic losses at $3 billion to $5 billion.

The native people who lived in villages along
the coast and on nearby islands have had their
way of life disrupted. They lived a subsistence
lifestyle, taking most of their food directly from
the ocean and relying very little on outside
sources. But now, afraid to eat their usual diet,
they must find new ways to support themselves.
The disturbance of their centuries-long cultural
heritage has led to severe emotional problems
and social disruption.

One biologist, Kathryn Frost, remarked about
the environmental damage, "Emotionally, it is
overwhelming. . . . Old fishermen stand there just
sick. As a scientist, I'm supposed to be emotionally uninvolved—but I was heartsick." Many researchers believe it will take decades before
Prince William Sound recovers from the spill.

People all over the world were distraught by
the television and newspaper pictures and ac-
counts of the horror in Prince William Sound.
Many were angered even more by the way the
Exxon Corporation handled the disaster.

Cleanup after *Exxon Valdez*

The standby cleanup equipment and crews that
Exxon, one of the world's wealthiest companies,
had promised were nonexistent. Cleanup crews
did not begin full-scale work for three days, and
critical time was lost, allowing the pollution to
spread. The first efforts to contain the oil slick
were ineffectual. While company officials and
government agencies debated about how to clean
up the mess and who should take command, the
oil floated closer to the beaches, and more damage was done.

The cleanup crews tried using floating devices
called booms and skimmers to surround the oil

and clean it off the water's surface. They sprayed hot water to move the oil away from the beach. The hot water helped, but it also destroyed important microorganisms living between the rocks. They tried sorbents, materials that soak up oil, but they ran out of the material before much could be accomplished.

Volunteers came to the beach and tried cleaning oil-covered rocks by hand. But the size of the spill was overwhelming, and little protection was provided to the cleanup workers. Humans who inhale concentrated petroleum fumes risk damaging their lungs. Skin contact with oil can cause rashes and can even damage the human liver, kidneys, and bone marrow. After a few months, Exxon declared that the beaches were clean and ceased cleanup efforts. It spent $2.5 billion on the cleanup.

Clearly Exxon had not planned for this risk. A federal grand jury charged that Exxon had caused

Volunteers help clean a bird covered in oil. The disaster killed and injured thousands of birds.

the spill by knowingly employing unqualified crew members. In 1991, the company agreed to pay a fine of $100 million to settle criminal charges arising from the spill and another $1 billion to settle claims filed by the state and federal governments. When Exxon's chairman, Lawrence Rawl, commented publicly that these fines would "not have a significant effect on our earnings," U.S. District Judge H. Russel Holland in Anchorage rejected the settlement, as did the legislature of Alaska. The parties suing Exxon are determined to ask for larger amounts or to take the corporation to court for a trial.

Persian Gulf oil spill

While the world was still in shock over the *Exxon Valdez* spill and its aftermath, another, perhaps more shocking, ecological disaster involving oil and the oceans occurred. This time, it was a deliberately planned spill, an act of war.

In August 1990, Iraq took over its Middle Eastern neighbor, Kuwait, by military force, giving Iraq control of the vast Kuwaiti oil reserves. On January 17, 1991, the United States and several other nations took military action against Iraq and its leader, Saddam Hussein.

During the Persian Gulf War, Iraqi president Saddam Hussein ordered troops to damage a Kuwaiti oil pipeline and five supertankers. The resulting oil slick spread more than 250 square miles—the worst oil spill in history.

Around January 20, five supertankers filled with Kuwaiti oil were docked at the port of Mini al Ahmadi at the head of the Persian Gulf, the body of water bordered by Iraq, Kuwait, Iran, Saudi Arabia, and other important oil-producing nations. Suddenly these ships, along with a pipeline carrying crude oil from land storage tanks to Sea Island ten miles offshore, began to spill oil into the gulf. Hussein had ordered his troops to damage the ships and the pipeline, apparently in an effort to prevent a U.S. naval landing on Kuwaiti shores.

The ships continued to pour out oil for days. At least 140 million gallons of oil—more than ten times the amount spilled in Prince William Sound—flowed into gulf waters for at least seventy miles along the shores of Saudi Arabia. In the midst of war, nothing could be done to stop the flow of oil, and cleanup could not begin until fighting stopped in April 1991. Within a week the slick covered more than 250 square miles.

Contaminating the Persian Gulf

Labeled ecological terrorism, Hussein's act produced history's largest oil spill. The murky slick covered an area a dozen times larger than the *Exxon Valdez* disaster, thirty-five miles long and ten miles wide. Part of the oil slick was set on fire, emitting ash into the air that reached as far as Hawaii, eighty-seven hundred miles away. During the war, hundreds of land-based oil wells in Kuwait were also set ablaze.

The spilled oil threatened to shut down the overall economy of the area, which depends on

gulf water for cooling in industrial processes. It also crippled the gulf's multimillion-dollar shrimp and fishing industry.

Wildlife, including coral reefs, turtles, birds, dolphins, and ecologically fragile wetlands, suffered severe harm. The slick contaminated sea grass beds that provide food for marsh birds and ruined rare stands of mangrove trees.

Because the Persian Gulf is shallow compared to areas farther out in the ocean, only 110 feet deep on average, it takes two hundred years to flush out all the water in it. Natural cleansing has helped in Prince William Sound, which receives all new water every few days. In the Persian Gulf, oil has both washed ashore and sunk to the bottom. One environmentalist described the effect as like "paving the gulf in asphalt." As in Alaska, the oil companies were poorly prepared with equipment for containment or cleanup, even if they had been able to begin right away.

The legacy of an oil spill

The extent of the damage of the Persian Gulf spill is still unknown. Scientists have learned that the damage caused by an oil spill depends on many factors: the grade or type of oil spilled and the rate of speed it moves through the water, the temperature of the water, the physical features of the area, and the weather, wind, and ocean current at the time.

For instance, in Prince William Sound, the frigid temperatures of the water kept the oil from breaking down or biodegrading as quickly as it might have in warmer seas. Left untreated in the cold water, the oil congealed into a thick, batter-like paste that clings to living bodies, plants, rocks, and sand. Called mousse, and whipped into a froth by the strong winds, this paste is almost impossible to remove or clean.

It may take two hundred years for the Persian Gulf to cleanse itself of Kuwaiti oil.

On the other hand, warm waters, with the help of sunshine, begin to thin the oil out immediately. When oil thins and disperses, it does less immediate harm and eventually biodegrades. However, the ultraviolet sun rays in warm climates change the composition of the oil, making it even more toxic than before.

Any type of oil spilled in any type of water in large amounts injures all forms of life. Sea bottom plants and minute living organisms die. Fish gills are clogged and the fish smother. The baleen fibers in the mouths of whales are clogged. Some whales catch their food through these sieves. Unable to feed, they starve.

Oil also penetrates the fur of mammals and feathers of birds, weighing them down to drown or depriving them of insulation so that they die of

A Middle Eastern man walks along a Kuwaiti seashore polluted with oil.

A man examines a bird killed by an oil spill off Huntington Beach, California.

hypothermia. If they survive, they try to clean themselves and ingest the oil. Then their intestines become clogged with the toxic poisons, and they die.

Oil spills are a relatively new phenomenon, and scientists are just beginning to get an idea of the effects they have on the ocean environment after a few decades. Effects that last longer than that will not be known until more time has passed.

Long-term effects

The fact that oil is a natural substance and is biodegradable makes for disagreements among scientists as to the long-term effects of oil spills. Some believe that when oil is no longer visible on water surfaces, beaches, or rocks, the thinned-out oil will be biodegraded by natural bacterial organisms, and the remaining waste will be recycled into the ecosystems.

However, after studying past spills, more and more scientists agree that long-term damage is much more extensive and may continue for years. Collected data show that oil ingested by plankton is expelled to the ocean bottom. There it remains in the sediment, poisoning bottom fauna, or plants. As other species eat these plants, the oil recycles in the food chain. Each time it is ingested and expelled, it begins to work again in the bottom sediments. It is thought that this happens in both shallow and deep waters of the ocean.

Toxic oil kills ocean coral, some shellfish, lobsters, and other exotic warm-water species like starfish and sea urchins. Its poison kills all fish and their eggs and larvae. This loss of eggs and larvae broke the reproductive cycle of salmon and herring in the *Exxon Valdez* spill, and such broken cycles can deplete a species.

Tar balls of oil float with currents through waters around the world. Turtles swallow these balls and die as their intestines are clogged and poisoned.

Gravel and sand beaches may look clean on top, but they absorb oil into deep layers. Marine worms and minute creatures of the food chain dig into these sand layers, ingest the oil, and die. If they do survive, their reproductive cycles are often disrupted. Oil remaining on rock surfaces prevents many rock-clinging species from establishing new colonies.

Most scientists have no doubt that life will go on in badly damaged ocean areas after a long period of time. But they are not sure how long that time will be, and they warn that many species may be unable to adapt and may die out.

As scientists and corporations debate ways to deal with oil spills, one writer offers an answer. Says Gladwin Hill, "The only way to deal with oil spills is to prevent them."

5

The Debate over Deep Ocean Dumping

Most of the damage caused by ocean pollution has been measured in coastal areas. This makes sense for several reasons. It is easier to examine ecosystems in shallower water and in places where scientists know what kinds of waste materials have been dumped in the ocean. Also, investigators are naturally concerned about potential harm to human communities near the seacoasts.

But what happens to the deeper areas of the ocean as a result of pollution? How much do scientists really know about the effects of toxic chemicals, radwastes, oil spills, or sewage when they enter these locations far from the coasts? Does deep ocean dumping pose a serious threat to the environment, or is it the solution to the problems known about coastal dumping?

How deep is the ocean?

Continents, the largest land masses on earth, are surrounded by an area called the continental shelf. On the continental shelf, the floor of the

(opposite page) Scientists want to learn more about the effects of deep sea dumping.

77

ocean, or seabed, slopes away from the land very gradually. The size of the continental shelf varies throughout the world, but it is typically less than one-tenth of a mile deep, about four hundred feet. This shallower area may extend for only a few feet or as much as 500 miles; on average it extends about 43.5 miles out from the seacoasts.

Beyond this shelf the ocean becomes much deeper. Oceanographers estimate that the average depth of the ocean is two-and-a-half to three miles. The deepest areas, believed to be six to seven miles in depth, are known as the abyssal plains, but they are not smooth. Scientists have found that the abyssal plains in the center of the oceans have huge ridges or mountain areas, some as high as mountains on earth. The abyssal waters also contain deep depressions or basins.

Scientists disagree strongly about the results of dumping polluted materials into the abyssal parts of the oceans. Some believe that the seas, covering more than two-thirds of the earth's surface, are so vast that they are the only place people can put the tons of waste materials created every day. As long as toxic materials are kept away from coastal areas and known wildlife ecosystems, according to this opinion, they cannot harm any life.

Other scientists believe that dumping in the deep ocean will damage ecosystems there that we know little about. They say that eventually that harm will spread to the ecosystems closer to land and will affect human life.

The capacity of the oceans to absorb waste materials is a matter of continuing debate among oceanographers. The questions raised by this discussion are not likely to be answered soon because not enough information is yet available. Science is still learning how to examine and measure the effects of deep ocean dumping.

The argument for deep ocean dumping

Almost all scientists agree that the waters near coasts are the most fragile of all ocean environments, and this is where change does the greatest harm. Oceanographers have found that the largest undersea communities thrive and grow near the land. These waters are shallow enough for plants and animals to be reached by sunlight, which is vital for their growth and survival.

Delicate undersea communities need clean water and unrestricted sunlight to survive.

More than half the planet is covered with seawater more than two miles deep. The seafood people eat comes from the top mile, and most of it is caught within two hundred miles of land. Thus, humans use only 2 to 3 percent of the earth's total of 300 million cubic miles of seawater for fishing. Businesses also acquire needed supplies of sand, gravel, oil, and gas from these coastal waters.

Most U.S. coastal waters have already been damaged to some degree by human waste disposal. One response to awareness of this pollution has been to put municipal, industrial, and radioactive wastes on ships that sail far out into the ocean and pipe them directly into water there. The theory behind this approach is that coastal areas can be protected and deep ocean areas used for dumping with little or no resulting harm to any ecosystems. These deep areas are also inviting for the cleaning of oil tank holds.

The ocean abyss

The middle of the Atlantic and Pacific oceans contain abyssal hills and plains that some scientists compare to deserts. Marine life there is sparse, and mineral wealth nonexistent. Thus, some believe deep ocean offers the greatest potential for low-risk waste management.

The disposal of radwastes is a typical problem. At present, U.S. nuclear power plants store their low-level wastes on their own properties and bury high-level wastes in large underground steel tanks on military sites or in deep earth deposits of salt and clay. But these solutions will not work for long. The nuclear power industry throughout the world is accumulating billions of tons of radioactive wastes, and land areas with adequate deposits of salt and clay to store them are limited. Some scientists suggest that radwastes and other

toxic materials can be safely buried in the seabeds in abyssal waters.

Charles Osterberg, a DOE (Department of Energy) oceanographer, suggests that laws should allow the oceans to be used for dumping of hazardous wastes. He argues that human beings spend little time on the deep ocean. The desire to protect this area is emotional, not based on rational experience and scientific knowledge, he says.

This argument also points out that if the most toxic wastes of civilization cannot be dumped in the ocean, they must be placed on the land where people live or in the air they breathe. These alternatives may be more dangerous than ocean disposal. And the capacity of land is limited. While more than 80 percent of U.S. waste is now disposed of in sixty-five hundred landfills, those will reach capacity in a few years. A landfill leaches into groundwater, the source of drinking water, and can contaminate it. At least two thousand of

As the amount of land available for landfills decreases, ocean dumping increases.

Some scientists believe the deep ocean is the only logical place for waste disposal.

these landfills will be closed by 1995 because they are full or no longer safe to use.

Some small states like New Jersey transport their municipal wastes to landfills in other states or ship them to developing countries to be buried. But these practices are on the decline because developing countries are becoming reluctant to receive the garbage of others. So, where are the wastes going to go?

Whose backyard?

When waste is dumped into the ocean, gravity probably carries the particles downward to the ocean bottom. Some scientists believe that it is foolish for human regulations to ban use of what appears to be a natural disposal system.

Charles D. Hollister, a senior scientist at the Woods Hole Oceanographic Institute, explains that human societies have four "backyards" into which to put waste—space, air, land, and sea. The use of space would be too expensive and would require the use of rocket fuels that would pollute the atmosphere. Much more harm is done to humans and other living creatures by landfills and air pollution than by ocean pollution, this argument goes.

So the sea is the only backyard that is feasible for waste disposal, Hollister suggests. With proper study and planning to minimize risk, he says, the oceans are the safest place to dump sludge wastes and biodegradable garbage. Hollister also suggests that if the right technology were developed, even very toxic heavy metals and radioactive materials could be disposed of safely in the ocean.

At the bottom of deep sea basins are vast underwater fields of clay mud resembling creamy peanut butter. Heavy metal atoms called ions are attracted to this mud and stick to its particles. This material, miles thick, is elastic and could hold drums of toxic wastes if they were dropped in the right way, Hollister believes. If waste leaked from those containers, gravity would probably keep it down. He emphasizes that more experiments and research are needed to employ this disposal technique safely.

An immense portion of the central North Pacific is more than three miles deep. For the last sixty-five million years, mud has formed in this area. There have been no earthquakes or other natural activity, and few signs of marine life have been seen. So no geologic catastrophe is likely to occur for millions of years, scientists say. Therefore, it would be safe to bury dangerous wastes in these parts of the ocean.

While some scientists believe that even the most dangerous waste can be disposed of in the deep ocean, with the proper precautions, others take just the opposite view. They say that because so little is known about the ecosystems of the deeper parts of the ocean, it would be foolish to risk dumping any waste there.

The argument against deep ocean dumping

Many scientists who study the environment believe it is unwise to assume that dumping even nontoxic or nonhazardous wastes in deep ocean zones is a safe practice. At one time, they argue, it was assumed that nuclear fallout, minor oil spills, the burning of coal, or the use of insecticides such as DDT would do little harm. All of these practices have proved disastrous to the ecosystems and to human life.

Another concern is the growing interest in mining some parts of the deep ocean floors for important minerals. If not carefully monitored, this activity could cause new forms of pollution such as metal poisoning of species in the deep ocean.

Cleanup is much more difficult than preven-

A load of sludge fouls the waters off the New York coast.

tion, goes this argument. Even though coastal areas are being cleaned up and new laws severely limit waste disposal there, cessation of dumping will not reverse the degradation of the shoreline ecosystems for some time.

Dumping in the ocean does not always mean the material will sink directly to the bottom. Ocean currents are powerful and may carry materials for miles and to shallower water. Because of the slowness of the deep circulation, serious and irreversible harm to the ocean ecosystem could be inflicted before the effects become noticeable.

Measured harm

Pollution has already been discovered in deep ocean areas. Because of this, some countries have banned all ocean dumping. Anne W. Simon, an author on environmental topics, writes that dumping anywhere in the ocean may kill plankton and thus reduce the ocean's productivity. Some scientists say that deep ocean plankton are more vulnerable to pollution than their inshore cousins, which have adapted somewhat to new materials in their environment.

Some evidence of harm done in deeper ocean waters comes from examination of the New York bight, 106 miles out to sea, where a company called American Cyanamid dumped a million liters of chemical waste. Marine scientists at Woods Hole cloned, or genetically reproduced, some plankton from the coast and other plankton from the deep water. They placed both samples in the same polluted solution and compared their growth. Test results indicated that deep ocean and coastal plants are equally susceptible to pollution damage.

Powerful effects

The ocean explorer Jacques Cousteau points to the evidence of how far pollutants have been carried by the oceans as an indication that wastes dumped far out at sea could still harm life on and near land. Because of the way water moves, DDT has been found in the livers of penguins in the Antarctic where there is no pollution. In 1985, Cousteau wrote:

> In 90 years there will not be one drop of water in the Mediterranean that is there today. The pollutants in that sea will finally come to pollute the rest of the oceans. The same is true for the Caribbean, the North Sea, the Gulf of Finland and so on. While rivers and enclosed or semi-enclosed seas are in worse shape today than the open ocean, that may not be true in 10 or 20 years.

Those who oppose all ocean dumping argue that even small amounts of pollutants, especially toxic ones, could have powerful effects on the world's ecosystems. Simon notes that tiny amounts of DDT, PCBs, and radioactive fallout have been detected in the deepest parts of the ocean. "Their presence in the deep sea, no matter how small the amount, means we have changed

Ocean explorer Jacques Cousteau believes that waste dumped far out at sea can still harm life near and on land.

the ocean for as close to forever as man can measure," she said.

In tests of some benthic fish, those that live at the very bottom of the ocean, traces of chemical pollutants have been found in their livers. This deep ocean environment has been constant for thousands of years, Simon explains, but changes have already been caused by toxins that have existed for only a few decades. Those who oppose all ocean dumping believe there is no threshold of safety and that the concept that the ocean assimilates or absorbs materials added to it is absurd.

Solutions to the dilemma

Most scientists would agree that questions about pollution of deep ocean areas cannot be answered until more information has been gathered. For example, oceanographers have discovered that in deep parts of the Atlantic Ocean, where the water is usually very still, blizzardlike storms strong enough to lift sediment from the seafloor sometimes occur. These storms stir up the bottom, pick up mud, and distribute it downstream into mounds more than 480 miles across. The biggest supertanker full of sewage sludge could hold only one-thousandth of the mud carried by a single benthic storm. Scientists must find out if it is possible to predict when and where they will occur so that these areas can be avoided if deep ocean dumping is to be done.

Oceanographers know how masses of water move and how materials are transported. But scientists say that more study is needed of how wastes are distributed over space and time and of the damage caused by toxic accumulation over long periods of time. Then they will know if technology and monitoring systems can be developed that would permit safe dumping in the ocean.

Thorough study would enable waste disposal

policies to be based on better selection, management, and monitoring of sites. After the lessons learned from pollution of coastal waters by mercury, oil, and other substances, many scientists encourage continuous monitoring. One reason these occurrences were so disastrous is that no one was measuring the ongoing physical, chemical, and biological effects of these wastes until a problem arose.

Help from technology

Another need is for standardized measuring equipment and methods. Fisheries and meteorological stations use research boats, buoys containing instruments, airplanes, and satellites to measure wildlife activity and pollution in the ocean. With the use of such technology as infrared instruments, which measure wavelengths longer than those of visible light, remote ocean areas could be monitored more closely. If scientists could establish standards for measuring the dangers posed by various amounts of pollutants, everyone involved—science, business, government, the public—could talk more intelligently about ocean pollution and find solutions.

At this time, no U.S. government agency is charged with the responsibility of exploring the option of waste disposal in the ocean. No federal funding has been allocated for research on the environmental effects of deep ocean dumping or on technology for waste disposal and monitoring. To make intelligent policies and laws, such support is needed. Otherwise, humanity will learn about deep ocean pollution the way it learned about coastal pollution—by examining the damage after tragedies have occurred instead of preventing further harm to plants, animals, and humans.

Using specialized instruments like this infrared camera, scientists can more closely monitor remote ocean areas for signs of pollution and waste disposal.

6

Saving the Oceans

REPORTS OF BEACH closings and warnings about seafood contamination in the early 1980s frightened many people. Throughout the nation, Americans became greatly concerned about the declining quality of their coastline environments. In response to this concern, the National Oceanographic and Atmospheric Administration (NOAA) created a six-year research project. Under management of Thomas P. O'Connor, the National Status and Trends (NS&T) program began in 1984.

From eastern to western shores, three hundred sites were chosen to represent the urban estuaries and coastal waters of the nation. The sites selected were mostly near communities of about 100,000 people, not near huge cities or places of heavy waste discharge.

Sites containing large colonies of mussels and oysters were of special importance. The NS&T project was designed to evaluate long-term effects of human-dumped wastes on various species of fish and shellfish. Mussels and oysters are good indicators of the contents of ocean sediments. They can survive poisonous contaminants, and they store sediment content in their tissues

(opposite page) Volunteers take to the beach in an effort to clean the shore.

and muscles in amounts that can be easily measured.

The NS&T study sought information about three major questions: the spread or distribution of certain chemicals and metals; the amounts, if any, of substances in the bodies of fish and shellfish; and the changes, if any, in concentrations in sediments of PCBs, DDT, and a few other toxic industrial wastes.

The project's results were published in 1990. The researchers found that many areas are clear or nearly clear of waste contamination. They discovered that former high concentrations of PCBs and DDT have diminished substantially in some areas. This may have occurred because these chlorinated compounds were banned from production many years ago and have not been dumped in U.S. waters for ten or more years.

Toxic pollution has affected oceans around the globe. Here a four-man crew attempts to rig a floating boom to suppress the flow of oil from the stricken tanker Torrey Canyon.

However, DDT is still being used in other parts of the world, and PCB products are in use in all nations.

This overall reduction of toxic pollution in the coastal environment indicates some hopeful possibilities. Several new laws have been enacted in the United States and other countries. Fines and restrictions are imposed on anyone who does not comply. These regulations seem to be leading to improvements in the conditions of coastal ocean waters throughout the world.

Laws slowly take effect

Before 1960, few people were concerned over ocean waste dumping. The oil spill from the *Torrey Canyon* in England in 1967 sparked the first public interest in protecting ocean environments. Marine scientists began to publish reports warning that oil spills and waste dumping had affected marine life and ecosystems badly. An important publication in the United States was the Council on Environmental Quality's report in 1970.

That year the United States and other industrialized countries began to evaluate the effects of ocean dumping. As a result, during the last two decades, most western countries have made some attempt to regulate and control ocean dumping.

In 1972, the U.S. Congress passed the Federal Water Pollution Control Act, also known as the Clean Water Act, and appointed the Environmental Protection Agency (EPA) as the governing agency over all dumping of wastes in U.S. waters. The Coast Guard was assigned to enforce these regulations. All ocean dumping would require an EPA permit. Outfall dumping of untreated wastes was banned by a later amendment. Under the U.S. Marine Protection, Research and Sanctuaries Act, or Ocean Dumping Act, passed the same year, any material that would adversely affect human health, welfare, the marine environment, ecosystems, or economic interests was banned from disposal in the oceans.

Also in 1972, thirty-three nations met and signed a treaty called the London Dumping Convention (LDC). Following U.S. guidelines, this treaty forbade ocean dumping of several harmful substances, including mercury and cadmium compounds, some plastics, oil deliberately dumped by ships, high-level radioactive wastes, and biological and chemical warfare materials. Permits must be obtained for dumping of other wastes in international waters.

More treaties to prevent pollution

By December 1989, the LDC had been ratified by sixty-four nations. According to the LDC's records of permits issued, the amount of waste material dumped into the oceans rose steadily from about 125 million tons in 1976 to about 275 million tons in 1985. However, most of this waste was dredged material, solid matter produced by digging into the earth. The amount of industrial waste and sewage sludge remained about the same.

Two organizations were formed in the early 1970s to monitor wastes dumped by ships at sea and to check the condition of ships. The International Convention for the Prevention of Pollution from ships, or MARPOL-73, was created in 1973. MARPOL has been further amended with tighter restrictions in the years since. The Convention for Safety of Life at Sea, or SOLAS-74, was formed in 1974.

Throughout the 1970s and 1980s, nations endeavored to slow the deluge of wastes entering the oceans by forming a number of other national and international regulatory agencies. One is the United Nations Law of the Sea Conference,

The amount of garbage and waste materials has risen steadily over the years.

People's carelessness has caused much unnecessary ocean pollution.

which convenes regularly to develop a treaty for management of the oceans.

But despite the good intentions of these actions, a long and difficult time was to come. Although many laws were put into effect, industries, cities, and nations broke them whenever they deemed it necessary. Many of the new agencies lacked the money or staff to carry out their policing duties. Bribery, politics, and bad management often interfered with enforcement of the laws. Ocean waste dumping continued and even increased.

During the 1970s and 1980s, the world saw more oil spills, more dead and dying marine animals washed up on beaches, and more contaminated fish catches. Human cancer cases increased in areas near industrial discharges. Rarely a week or month passed without a newspaper story about an incident of environmental damage resulting from ocean pollution.

Citizens of many countries joined together in several environmental protection organizations, which brought these problems to the attention of the public. The resulting public outcries brought pressure upon governments to enforce existing laws. The tables gradually turned.

Now, the United States spends more than $130 million per year on coastal environmental monitoring. Other countries and international groups are taking serious steps to enforce compliance with laws protecting the oceans.

Changes around the globe

Today, international waters are policed under the MARPOL agreement. Ships at sea can no longer dump their wastes into the ocean. To be sure they do not cheat, all ships are officially checked when they dock at seaports. Their wastes must be properly bagged and ready to be taken ashore for disposal in legal land sites. If a ship docks with no wastes, it must have legal records of how they were disposed of and the amounts involved. Some modern ships have incinerators or garbage-compacting containers.

Ships that break these rules are denied future entry into the seaports of the sixty-four LDC nations. If a ship cannot enter these major seaports, its operators cannot make profits.

Both the Coast Guard and the United Nations are investigating ways to require improved design of oil tankers and better training of tanker crews in an effort to eliminate oil spills. Another international effort involves plastics.

In 1984, NOAA founded the Marine Entanglement Research Program to push for ratification of an international ban on ocean dumping of plastics. That was accomplished when fifty-two countries signed a 1988 amendment to MARPOL. It is enforced in the United States by the Coast Guard and the U.S. Department of Agriculture, both of which have reported improvements in the first two years. However, military vessels are exempt.

Many countries are enacting their own laws on ocean pollution. The European Economic Community (EEC), which includes many nations, is forming domestic rules on fouling of its waters, as is Japan. In 1970, the United States ceased dumping radwastes into ocean environments. Japan stopped soon after, and a recent international agreement banned ocean dumping of all nuclear wastes.

Great Britain has announced it will phase out dumping of industrial wastes by 1993 and sludge disposal in the ocean by 1998. A U.S. ban on dumping sludge into the oceans goes into effect on December 31, 1991. New York City has said it will keep dumping until June 30, 1992, and pay the fines. Southern California's outfall will continue for awhile until new systems can be developed.

With federal governments and international

The plastics industry has agreed to help recycle plastics. This man points to the types of recyclable containers.

treaties setting the pace, cities and industries are also responding to the cry for cleaner oceans. Some U.S. industries say they want to make up for their past pollution of the ocean, and most have been restricted by law from continuing their harmful practices.

The plastics industry has pledged to help recycle plastics produced in the past and remove them from the environment by 1995. A new, safe, biodegradable type of plastic is being developed, based on vegetable starches. While it may not be as strong and durable as synthetic plastic, it will be serviceable and harmless to the environment, according to the plastics industry.

Chemical companies now must meet standards for clean wastes imposed by the EPA. Those who belong to the Chemical Manufacturers Association (CMA) are required to follow CMA standards, which exceed those of the EPA. Any chemical manufacturers that refuse to comply will be expelled from the CMA.

One reason industries are becoming so cooperative after decades of disregard for the environment is that it is now economically profitable to prevent pollution. Consumers, now increasingly aware of the damage done to the environment, may not patronize companies that pollute. Some manufacturers have discovered they can cut costs by running cleaner operations. They avoid fines this way and it is less expensive to monitor dangerous materials before they are used than to clean up their poisons afterwards. And workers who face fewer dangers are less likely to bring lawsuits or create damaging publicity.

Recycling by industry

A new "trading wastes" industry—recycling on a large scale—has begun. The EPA plans to create a data base of waste products that can be

Paper, plastics, and many other kinds of materials can be recycled.

traded among industries. For instance, the rubber scraps from a tire maker could be used to make highway asphalt. A company with excess cat litter can sell it to animal shelters or to highway departments to be used to soak up oil spilled in road accidents.

Even toxic wastes can be salvaged. Sludge often contains metals that can be extracted and used again. In the past, industries considered this process too expensive to bother with. But today, many metal sources are becoming depleted, and industries are relying more on reuse of metals from waste. As a result of more emphasis on recycling, some people believe, fewer harmful pollutants will be dumped in the oceans.

Although progress in cleaning up damaged coastal ecosystems is apparent, there are still unresolved issues relating to outfalls, pollution that reaches the ocean through rivers, catastrophic oil spills, and exploited coastal regions. Laws alone will not ensure clean oceans in the future. More study, publicity, and personal commitment are essential to bring about changes in people's attitudes about the use of resources.

Ordinary citizens, small communities, and concerned groups have had an impact on efforts to save the oceans in many ways. Sometimes they get involved in specific projects as volunteers.

Intensive annual beach cleanups are held in New Jersey and New York. In 1986, the Center for Environmental Education (CEE) organized a Texas Coastal Cleanup Campaign. The goals were to educate people about the problems caused by marine debris and to collect information about the trash found on the beach. Some twenty-eight hundred volunteers, including school-age children, participated at twelve coastal sites. They filled seventy-nine hundred trash bags with nearly 124 tons of debris gathered over a distance of 122 miles. The most common item was plastic bottles—16,572 in all. Now Texas has an Adopt-a-Beach program in which groups can commit to being responsible for keeping one beach area clean.

During major oil spills, volunteers work to-

Recycling metals and other kinds of materials means less dumping of wastes into oceans.

gether to clean oil from birds, animals, and rocks. Most have learned firsthand about the devastation of oil spills when they see that their efforts have little impact in the face of all the damage.

In households that use products containing synthetic compounds, people are gradually becoming more aware of what happens to the waste. It is important to dispose of only organic materials compatible with the marine environment in the ocean. People should not discard waste near areas where others use the beaches for picnicking and swimming.

Recycling is an important local and individual effort. Almost everything can be reused, and most communities have recycling programs for glass, metal, paper, plastics, and other materials.

People from all walks of life are active in environmental groups that try to educate the public about pollution and spread word about how it can be prevented. These groups emphasize a lifestyle in which people use smaller amounts of the earth's resources, such as oil for energy, thereby producing less waste. They also conduct studies and give testimony when new laws and regulations are being considered at local, state, and national levels.

Many people are volunteering to help clean up the beaches and shores, ridding them of unsightly garbage and waste materials.

Alternate means of disposal

While the debate about dumping of wastes in the deep ocean goes on and while people try to clean up the harm already done by pollution of coastal areas, more waste is produced every day. People continue to create garbage, industrial wastes, oil, sewage, medical wastes, plastics, radioactive materials, heavy metal compounds, chemical compounds, and other pollutants. Since it is not likely that human societies will completely stop producing waste, they must design waste management programs that minimize risks to human health and the environment.

Some alternatives to ocean dumping exist, but they may not solve all the problems of potential pollution. Leakage is a problem in landfills and in containers used to bury toxic materials. Incineration of wastes may destroy the hazardous elements, but it pollutes the air, creates toxic ash, and requires expensive technology. Chemical neutralization reduces toxicity but still produces waste. More investigation of alternatives is needed.

Scientific support

In the past, the scientific community has been absent at decision-making time on ocean pollution regulations. Scientists have published their opinions and findings but have not participated in setting public policies. When the Ocean Dumping Act of 1988 was passed, there was no unified scientific opinion on the issues it addressed. This act clarified and tightened restrictions from the earlier act. It banned dumping of sludge and radwastes except for research purposes. It was enacted by Congress on the basis of public debate, which may express great concern but does not al-

It will take much effort to reduce the waste materials found in the oceans and to protect sea life from the effects of these materials.

ways include complete information on the pros and cons of various forms of waste disposal.

Thomas R. Kitsos and Joan M. Bondareff, staff members of the Merchant Marine and Fisheries Committee of the U.S. Congress, would like to see more scientists testify at congressional meetings involving legislation on ocean activities. In one report they wrote, "In our democratic system of government, when the public demands environmental protection, and the scientific community fails to speak with one voice, Congress generally reacts by passing legislation to afford that protection." Input from scientists would help governmental bodies create policies and laws that are more reasonable, economical, and safe in the long run, not just based on human feelings.

Decisions about waste disposal affect the future of the human race, not just the oceans. These decisions need the expertise of specialists, including scientists from many fields. Judith E. McDowell Capuzzo, senior biologist at the Woods Hole Oceanographic Institute, has written, "As we approach the 21st Century, it is essential that scientists, environmental managers, policymakers, engineers, and legislators, work together to develop environmentally sound waste-disposal options." It will take the unification of all types of human knowledge and endeavor to create a future that allows safe, productive oceans.

Glossary

abyss: Area of the ocean beyond the continental shelf.

bioaccumulative: Increased amounts of a substance or element that build up in living species.

biodegradable: Capable of being broken down by natural decaying processes.

carcinogen: Substance that produces or tends to produce cancer.

chlorinated compound: Chemical compound made by combining chlorine and petroleum.

coliform: A type of raw sewage bacterium.

continental shelf: The gently sloping portion of a continent that is submerged in the ocean, resulting in a rim of shallow water surrounding the landmass; the shelf's outer edge drops steeply to the deep, oceanic abyss.

dispersal: Breakup of the particles of pollutants into tiny bits.

ecosystem: A community of living plants and animals that depend on each other for life.

groundwater: Water within the earth that supplies wells and springs.

heavy metal: Metal that is at least five times heavier than water.

high-level nuclear waste: Nuclear waste that remains radioactive, and thus hazardous, for hundreds to thousands of years.

industrial waste: Waste resulting from the manufacturing of various products; often contains chemicals and heavy metals.

landfill: A legal garbage dump in which waste is buried between layers of earth in low-lying land.

low-level nuclear waste: Nuclear waste that remains radioactive, and thus hazardous, for hours, days, weeks, or months.

marine biology: Study of the plant and animal life of the sea.

mousse: Oil that has formed into solid globs or layers of pasty ooze in cold temperatures.

oceanography: The science of the study of oceans.

outfall: Pipe that drains waste directly into the ocean.

PCB: Combination of various chlorinated chemicals and petroleum.

pesticide: Poisonous chemical and metal compound used to kill insects and other crop-damaging pests.

photosynthesis: The formation of energy when plants are exposed to light.

plankton: Minute organisms that live in sea water; plant plankton are called phytoplankton and animal plankton are called zooplankton.

plastic: A nonbiodegradable synthetic substance manufactured in many forms.

radioactive waste: The nonproductive by-products of a nuclear reaction.

radwaste: Another term for radioactive waste.

sediment: A solid material that settles in a liquid.

sewage: Waste material carried off by sewers.

sludge: The resulting solid matter after raw sewage has been processed and cleaned with water.

synthetic compound: Chemical or metal substance created by combining two or more natural elements.

toxic: Poisonous.

toxic ash: Poisonous waste that results from incineration of some materials.

trade waste: Another term for industrial waste.

Organizations to Contact

The following organizations conduct a variety of activities including research, educational outreach, and public awareness programs.

Oceanic Society
1536 16th St. NW
Washington, DC 20036
(202) 328-0098

The Oceanic Society is dedicated to the protection and wise use of the oceans and marine environment. The society conducts public education and awareness programs and conservation activities.

The Cousteau Society, Inc.
930 W. 21st St.
Norfolk, VA 23517
(804) 627-1144

The Cousteau Society is an environmental education organization dedicated to the protection and improvement of the quality of life. The organization conducts research and coordinates public activities. It publishes several magazines including *Calypso Log*, *Calypso Dispatch*, and *Dolphin Log*.

Greenpeace
1436 U St. NW
Washington, DC 20009
(202) 462-1177

Greenpeace opposes nuclear energy and the use of toxins and supports ocean and wildlife preservation. It encourages active, though nonviolent, measures to aid endangered species such as whales. The organization also monitors the global hazardous waste trade. It publishes the magazine *Greenpeace* as well as many books.

The Humane Society of the United States
2100 L St. NW
Washington, DC 20037
(202) 452-1100

The Humane Society works to end cruelty and harmful practices that threaten the well-being of animals, including marine mammals. The society publishes *The Humane Society News* and *The Humane Society Newsletter*.

The International Oceanographic Foundation (IOF)
4600 Rickenbacker Causeway
PO Box 499900
Miami, FL 33149-9900
(305) 361-4888

The foundation encourages scientific study and exploration of the oceans including the study of game and food fish, ocean currents, and the geology, chemistry, and physics of the sea. The foundation publishes *Sea Frontiers* magazine.

Suggestions for Further Reading

Charles E. Cobb, Jr., "Living With Radiation," *National Geographic*, April, 1988.

Barry Commoner, *The Closing Circle*. New York: Beekman Publishers, 1971.

Encyclopaedia Britannica. "The Ocean: Mankind's Last Frontier." New York: Bantam Books, 1978.

Don Groves, *The Ocean*. New York: John Wiley & Sons, Inc., 1989.

Don Nardo, *Oil Spills*. San Diego: Lucent Books, 1991.

Anne W. Simon, *Neptune's Revenge*. New York: Franklin Watts, 1984.

Bruce Stutz, "Last Summer at the Jersey Shore," *Oceans*, July/August, 1988.

Gilbert L. Voss, *Oceanography*. New York: Golden Press, 1972.

Works Consulted

Julie S. Bach and Lynn Hall, eds., *The Environmental Crisis: Opposing Viewpoints.* San Diego: Greenhaven Press, 1986.

Sharon Begley, "One Deal That Was Too Good for Exxon," *Newsweek*, May 6, 1991.

Sharon Begley, "Saddam's Ecoterror," *Newsweek*, February 4, 1991.

David K. Bulloch, *The Wasted Ocean.* New York: Lyons & Burford, 1989.

Judith E. McDowell Capuzzo, "Effects of Wastes on Oceans: The Coastal Example," *Oceanus*, June, 1990.

Jacques-Yves Cousteau and the staff of the Cousteau Society, *The Cousteau Almanac.* New York: Doubleday & Company, 1981.

Iver W. Duedall, "A Brief History of Ocean Disposal," *Oceanus*, June 1990.

Bryan Hodgson, "Alaska's Big Spill," *National Geographic*, January 1990.

Charles D. Hollister, "Options for Waste: Space, Land, or Sea?" *Oceanus*, June 1990.

Thomas R. Kitsos and Joan M. Bondareff, "Congress and Waste Disposal at Sea," *Oceanus*, June 1990.

Douglas B. Lee, "Tragedy in Alaska Waters," *National Geographic*, August 1989.

Thomas P. O'Connor, *Coastal Environmental Quality in the United States, 1990: Chemical Contamination in Sediment and Tissues*. Rockville, MD: National Oceanic and Atmospheric Administration, 1990.

Patricia A. Parker, "The Plastics Threat," *Sea Frontiers*, March/April 1990.

C.V. Reynolds, "Beachless Summer," *Discover*, January 1989.

Bruce R. Rosendahl, "Crow's Nest," *Sea Frontiers*, March/April 1990.

Derek W. Spencer, "The Ocean and Waste Management," *Oceanus*, June 1990.

Anastasia Toufexis, "The Dirty Seas," *Time*, August 1, 1988.

Michael Weisskopf, "Plastic Reaps a Grim Harvest in the Oceans of the World," *Smithsonian*, March 1988.

Index

About the Author

Maria Talen's creative experiences have been unusually rich and varied. She majored in drama in college, where she studied with playwright William Inge and the legendary stage actress Maude Adams. During World War II, Talen toured the South Pacific with the United Service Organizations (USO), which assist and entertain troops. Later, as a singer and comedienne, she performed in theaters and nightclubs throughout the United States. Eventually, she formed her own theatrical repertory company, serving as producer, director, and lead actress.

In the 1960s, Talen studied marine biology and conchology, the study of seashells. For the past nineteen years, she has made her home on Cape Cod, Massachusetts, where her handcrafted seashell artifacts and jewelry are well known.

Picture Credits